New Perspectives:
Renovated houses

Author: Arian Mostaedi
Publisher: Carles Broto
Editorial Coordinator: Jacobo Krauel, Cristina Soler
Architectural Adviser: Eduard Malgosa
Graphic Design: Oriol Vallès Garcia
Text: contributed by the architects

Edition 2007

© Carles Broto i Comerma
Jonqueres, 10, 1-5
08003 Barcelona, Spain
Tel.: +34 93 301 21 99
 Fax: +34-93-301 00 21
E-mail: info@linksbooks.net
www. linksbooks.net

All rights reserved. No part of this book may be used or reproduced in any manner whatsoever without written permission except in the case of brief quotations embodied in critical articles and reviews.

New Perspectives: Renovated houses

Index

8 **Franco Pinazza & Dieter Schwarz.** *Studio Units in Ennetbaden*

18 **Ramón Esteve.** *A terraced single-family dwelling*

28 **Mauro Galantino & Federico Poli (Studio 3).** *House in Orta Lake*

36 **Jean-Paul Bonnemaison.** *House in Lubéron*

46 **Ignacio Capitán.** *Conversion of an old clothing factory*

56 **Arnaud Goujon Architecte DPLG.** *Transformed penthouse*

66 **Albori Associati (E. Almagioni, G. Derella, F. Riva).** *House in Appenninos*

76 **Julian Cowie Architects.** *Fleetwood Place*

86 **Georges Maurios.** *Rue des Saint-Pères*

96 **Non Kitch Group bvba.** *Architecture and lifestyle*

106 **Architekturbüro Gasparin & Meier.** *Badehaus Ebenberger*

116 **Patrick Genard & Ass.** *House T*

128 **ARTEC Architekten.** *Raum Zita Kern*

138 **Klaus Sill & Jochen Keim.** *Apartment and Office Building*

146 **Jo Crepain Architect NV.** *Water-tower*

156 **Cristian Cirici & Carles Bassó.** *Vapor Llull*

170 **Fernando Távora.** *House in Pardelhas*

180 **Jean-Pierre Lévêque.** *House Rue Compans*

188 **Claesson, Koivisto, Rune.** *Private apartment / Town house*

200 **Littow architectes.** *Paris 6e*

208 **José Gigante.** *Wind Mill Reconverting*

220 **Studio Archea.** *House in Costa San Giorgio*

230 **Aneta Bulant Kamenova & Klaus Walizer.** *Conversion Sailer House*

246 **Fraser Brown MacKenna.** *Derbyshire St. Residence*

Introduction

Contemporary architecture is characterised by the use of new formulas and materials for creating inhabitable spaces in any conditions. Therefore, as a reaction to the current emphasis on new buildings, there seems to be an increasing interest in recovering buildings for conversion into new dwellings. When seeking a house or flat, an increasing number of clients prefer an alternative habitat that offers a similar degree of comfort but is filled with atmosphere and references to the past that fire the imagination. Until very few years ago, living in an old building meant enduring damp, leaks, cold and frequent repairs. People now want a home without these problems that satisfies all the needs of modern life, but they often still want a place with character in which to develop their lives. Due to the sharp increase in development and property speculation that has taken place in many parts of the world in the last few decades, a large number of old buildings have fallen into neglect and slowly deteriorated. These buildings reflect the passing of time and their walls indicate the different relationships that they have had with man. They form part of our landscape, and their disappearance would lead to an irreparable loss of our historical memory. A solution that allows them to be conserved is to convert them into sophisticated dwellings. The aim of this book is to show the main tendencies in the field of architectural rehabilitation through a varied selection of schemes by some of the most avant-garde and innovative contemporary architects in the last few years. Old flats, disused factories and warehouses, farms, a mill and even a water tower...any building, however old, can be converted into a modern space. New materials and the multiple solutions used in the schemes presented here give us an idea of the wide range of possibilities available for giving a new function to the built heritage. The complicated task of solving the technical and aesthetic problems, of building and conserving at the same time, is shown in the different forms of expression that are used, an intelligent architectural exercise in which respect and innovation are applied in the combination of new and old.

Franco Pinazza & Dieter Schwarz
Studio Units in Ennetbaden
S408 Ennetbaden, Switzerland

Photographs: Gaston Wicky, Zurigo

In the rehabilitation of this industrial building, the architects Franco Pinazza and Dieter Schwarz were able to transform an old "water workshop" into large apartments in which one can appreciate the modern spatial and volumetric configurations.

The building has seven bays. It is 35 m long, 17 m wide and 5 m high and has a warm colour and differentiated facades. On the main facade a cornice goes partially round the corner, whereas the north facade is blind and the south facade has two windows. The apartments are simple but show great attention to detail. They have a floor area of approximately 75 sqm, in addition to the gallery.

The dialogue between space and sculpture is based on three elements. The first is a gallery that acts as a compact element and is supported on two double "T" profiles that are painted blue. With a floor area of 16 sqm, this space is a perfect area for working or reading. The second element is a sanitary unit of stone located under the gallery, which includes a small kitchen, a WC and a shower. The walls are made of very dark bricks and the ceiling and floor are of dark artificial stone. Finally, between the sanitary block and the wall, there is a very transparent red staircase with treads of solid steel sheet treated with a non-slip material.

The materials were carefully chosen and the details seem refined but simple. The architects respected the existing elements and observed great care in dealing with them. They used relatively simple building work to evoke modern spatial and volumetric configurations that are not only aesthetic but also convincing and functional.

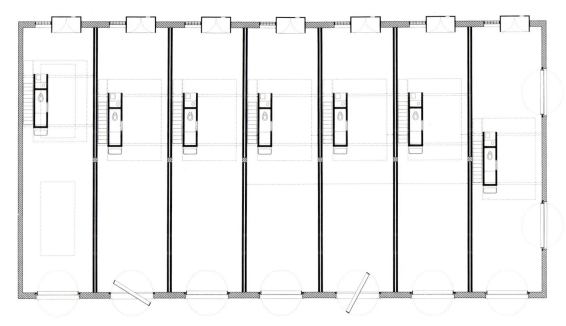

Floor plan

Pinazza and Schwarz intentionally break the rhythm at the entrance area of the west facade, facing the river, because the lines of the windows were repeated on the east facade. This new language achieves a new rhythm that enriches the reading of the walls of the entrances.

Ramón Esteve

A terraced single-family dwelling
Ontinyent, Spain

Photographs: Ramón Esteve

This project consisted in rehabilitating a terraced single-family dwelling located in an area that was originally developed outside the old town centre, but at the beginning of the 20th century it was the only remaining nucleus of the town.

The original layout of the house followed the lines of early urban dwellings for country people; a large front door giving direct access from the street, a narrow staircase, small and little-used inner courtyards, and a back garden that was used as a vegetable garden or to store farm implements. The restoration respected this initial layout and did not distort the essential structure of the dwelling. Practically all the secondary elements of the house were demolished, the walls were stripped, the floors were taken up, and all the floor slabs were demolished and reconstructed with the old wooden beams that were salvaged. The floor bricks were specially made of hand-made terracotta for the house. The bricks used to clad the courtyards and the floor tiles are made of hand-made terracotta, the doors and furniture of solid Iroko wood, and the external door and window frames are of zinc-plated steel.

The design is articulated on three floors with the following distribution: access from the street to the ground floor, housing the entrance hall, the garage, a cellar, an office, a toilet, the straight main staircase, and another open staircase that communicates the light well with the rear garden. The main staircase leads to the dining room, and from here a single step down gives access to the living room. The kitchen receives light from the garden through a french window that is also the exit to the garden. The garden is about the same size as the ground plan of the house, and its main feature is a lemon tree. From here, an external staircase leads to the second floor. A hall on this floor opens the way to two single bedrooms, two bathrooms, a dressing room and the main bedroom with a balcony looking over the garden.

In this scheme the traditional building techniques, rooted in the Mediterranean culture, adapt perfectly to the contemporary dwelling concept. The final result is an ordered combination of rays of light, clay, iron and wood.

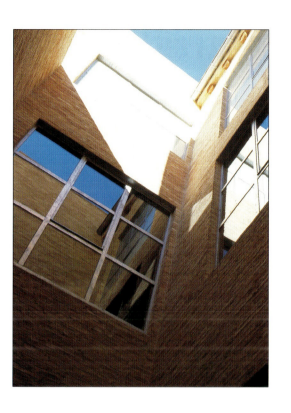

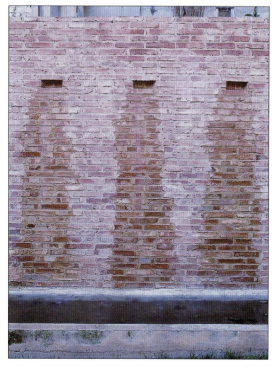

Mauro Galantino & Federico Poli (Studio 3)
House in Orta Lake
Orta S. Giulio, Italy

Photographs: Alberto Muciaccia

This unusual building is located in the Gothic district of an Italian town not far from Milan. The vertical nature of the house is conserved almost intact thanks to the two medieval walls that define the boundaries and the jetty.

Before the restoration, the building was in ruins, the ceilings were deteriorated and a large part of the foundations rested on the sandy bed of the lake. According to studies that have been made, this building was partly rebuilt in the 14th century, though the jetty was built in the 19th century. At first sight, it seems to be a simple rehabilitation: a productive residential microcosm. The domestic areas, such as the bedrooms and the living room, are organised vertically in the north "tower", while below the former cowshed, the henhouse, the garden and the jetty are organised with reference to the lake.

The restoration of this house -used as a second residence- respected the stipulations for the conservation of the cultural heritage with regard to volumes, walls and materials. The work was based on two objectives: to adapt the residential structure to new functions, and to obey the building regulations on the use of materials without sacrificing the possibility of a creating a new perception of the rehabilitated parts.

The result was a residential space composed of a living room of double height forming a horizontal, parallelepiped space with a covered jetty at the south end and a "tower" containing the living areas at the other end.

The jetty acts as an entrance door to this "residential microcosm". It was built in the 19th century to extend the horizontal volume of this medieval building.

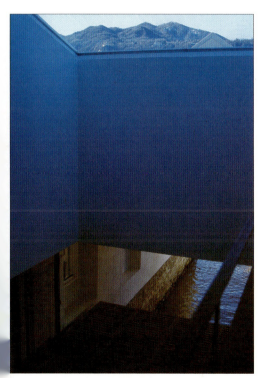

Jean-Paul Bonnemaison
House in Lubéron
Lubéron, France

Photographs: Leonard de Selva

The building is located strategically beside the church of a small village in Haute Provence, with good views of the street and the adjoining fields. The dwelling was built on the ruins of an old oil mill with the intention of preserving the memory of what was formerly one of the main sources of wealth of the area. It fits perfectly into the codes of this village, conserving several external walls and following a programme based on the urban volumetrics that amplifies the perspectives of the main street, thus respecting the style of the neighbouring buildings and the dominance of the church. The small, sharply sloping plot is fully occupied by terraces and gardens crossed by a path that leads to an existing swimming pool. The facades are clearly differentiated, and the public facade faces the town with five stone steps that emphasise the main entrance. The circular windows in the wall provide privacy and give the residents splendid views of the church steeple and the castle. This facade has a stone structure and has been intensified architecturally by means of jalousie windows that offer views of the town whilst protecting the interior from excessive sunlight. The other facade is private and totally glazed, facing the fields. Its design faithfully represents the changes of a society: new materials and construction systems combined with respect for nature.

The project is broken down horizontally into two groups, one containing the services and the guest area, and the other containing the main room of the house. Vertically, this main group consists of four levels, the last one with a bedroom, a bathroom and a terrace. The living room occupies three levels, with terraces that ensure spatial continuity with the sloping garden. A completely transparent metal staircase connects the different levels. The incorporation of this element on the glazed facade combines with the environment and allows the inhabitants to enjoy a different room, sitting in it as if it were a theatre.

The interior design is governed by the elegant and simple requirements of the clients, who are fascinated by Cistercian art, and includes hints of minimalism. The interior walls were painted white to capture the natural light of Provence and to highlight the collection of sketches that decorate the house. The light-coloured stone, the bluish grey of the structure and the beige of the cement areas fit into the general colouring of the town whilst giving the house a special character.

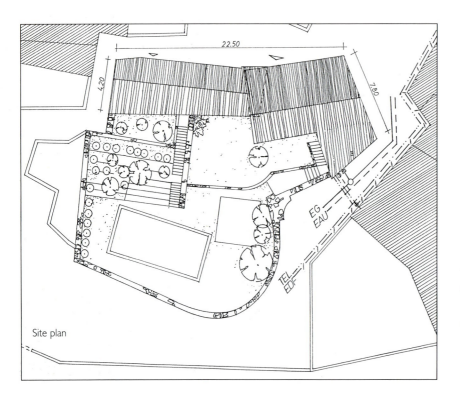

Site plan

The excellent situation of this dwelling and the absence of other buildings in front of the glazed facade allow the light to enter unhindered and give the landscape a dominant role.

The broken form of the staircase provides a continuous solution to the great difference in level between the floors of the dwelling. Its situation in front of the window converts it into a kind of a tier from which one can contemplate the landscape and the interior space.

Facade north-east

Ignacio Capitán
Conversion of an old clothing factory
Sevilla, Spain

Photographs: Ignacio Capitán

To look at the urban fabric of the Alcaicería (the old silk merchant's district) of Seville is to look through its pores. The interstices of a drowned city of rooms, commercial premises, galleries and other inhabitable cells, become desired "objects", boxes and corridors of natural light that are necessary for life and introduce a new disorder that transcends the immediate binomials: public-private, interior-exterior and street-house.

Rather than seeking the Muslim roots of this district, the new reading give itself up to the immediate, accepting past as present, yesterday as today and the heritage as a landscape for exploration and settlement. It is the result of the encounter between the new inhabitant and the proposed medium in which all the strata, all the histories, are superimposed, and new landscapes are formed.

In this project, the conversion of an old clothing factory into five dwellings and a shop, the block forms part of a continuous medium perforated by openings for light that make it inhabitable. It is therefore a building/block in which the main intervention consisted in capturing these openings of light by means of heaped boxes that furnish and structure the space that they encounter. It was thus possible to provide natural lighting and to make the block transparent: it can be sensed, but never intruded on. The glass of the boxes acts as a diffuser that controls the amount of light coming in whilst preventing crossed glances between the different properties.

The rest of the intervention was minimum, limited to designing spaces emerging through the roof that could enter into relation with the "figurative" nature of this landscape.

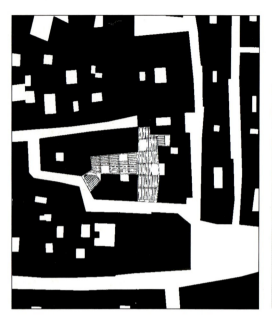

In the reconstruction of this block it had to be perforated in order to fill the interior with light and to create independent and inhabitable open spaces.

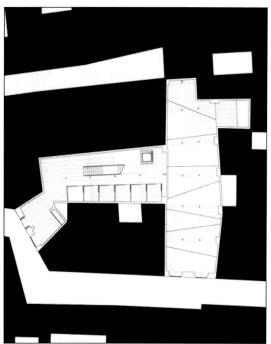

Ground floor plan

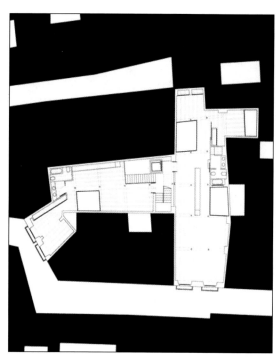

First floor plan

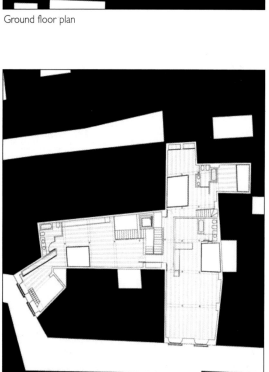

Second floor plan

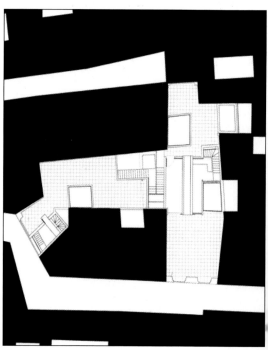

Third floor plan

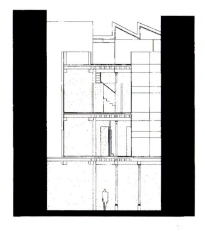 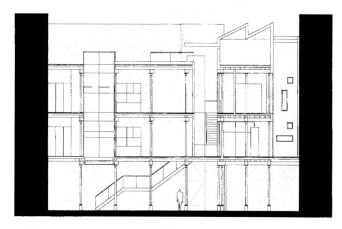

Section 2.2 Section 1.1

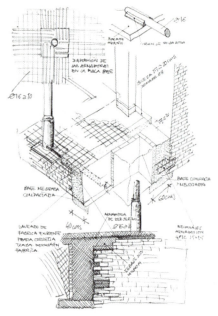

Foundation detail

Lighting through any surface, such as the ceiling or the glass boxes, is a constant that determines this whole work. The simplicity of the staircase of these single-family dwellings forms part of the detailed aesthetics of the scheme.

Arnaud Goujon Architecte DPLG
Transformed penthouse
Paris, France

Photographs: Joel Cariou

In the heart of Paris, the architect Arnaud Goujon transformed an old greenhouse located at the top of a block of flats into a small and comfortable refuge with a terrace and unique views. Conceived as an extension of the loft apartment, this volume would soon become the favourite room of this home. It is a scheme in which the initial volumetrics was respected and a new wooden frame was superimposed on the steel structure. On the exterior, the shingle boards are made of red Canadian cedar, while the interior walls are lined with moabi panels.

The main task for the architect in this rehabilitation - apart from the technical problems -consisted in designing and organising the different spaces of the apartment, and resolving the problems of execution and assembly of the different materials. The absence of exposed fittings on the wall panels of the interior helps to enlarge and unify the volume of the main room, which opens on both sides onto a terrace of 50 sqm covered with a jatoba wood deck and offering spectacular views of the urban landscape.

The interior of this unusual dwelling is composed only of a living room with an open integrated kitchen, in which a chimney is framed between two shelves, and a small bedroom with its bathroom. This room enjoys the benefit of two sources of natural light that illuminate this more private area: a small window in the back wall and a skylight located over the bed. The floor of the interior is made of chestnut parquet covered with white polyurethane paint that reduces the colour saturation and brings freshness to the dwelling.

The wood, chosen for its plastic and structural qualities, is used as a double skin: soft and beautiful in the interior and rough and sturdy on the exterior. Thus, although this organic material is set against the urban nature of an environment in which steel is the main component, its form fits well into the geometric pattern of the building.

The terrace running round this apartment is one of its fundamental elements. The panoramic views of Paris are an additional feature that enhances the architectural work.

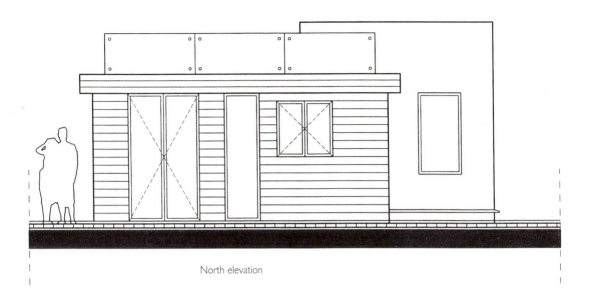

North elevation

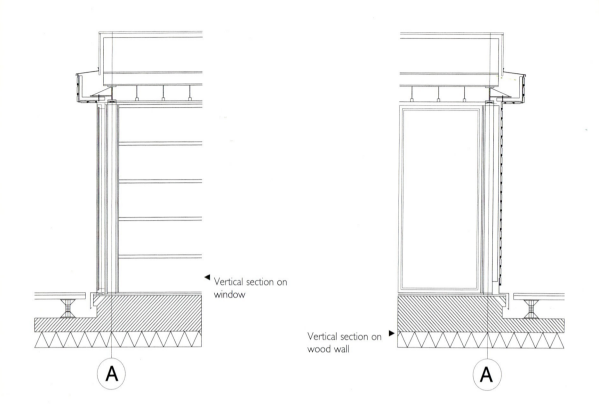

◀ Vertical section on window

Vertical section on wood wall ▶

Ⓐ

Ⓐ

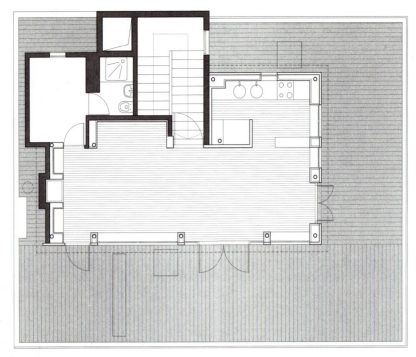

Floor plan

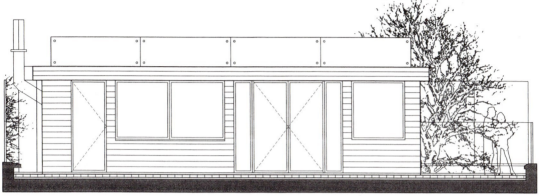

East elevation

Albori Associati (E. Almagioni, G. Derella, F. Riva)
House in Appenninos
Montese, Italy

Photographs: Matteo Piazza / Albori

This old farmhouse of an uncertain age, which has been remodelled several times, is set on a hillside of the Apennines in the region of Módena.

The idea of its new owners was to reduce the bathrooms and the night area as far as possible in order to leave the rest of the spaces open and at the disposal of the house. Because it was a building without partitions, in which the interior space coincided with the structure, most of the rooms were on different levels. This pure relationship between space and structures was so beautiful that it was decided not to alter it. Therefore, none of the rooms were divided and the new volumes are characterised by the type of material of the furniture inside them: wood, bricks and stone help to define the uses of the different rooms. The main barn had a certain majesty, with an irregular geometry, a floor area of forty-eight square metres and a roof that is 6 metres high at its highest point and 2.4 metres at its lowest point. After the demolition of the floor located between the ground floor and the first floor, a large room of double height perfectly illuminated by a long thin skylight was created. The darkest area in the dwelling was illuminated thanks to the courtyard located at the end of the old barn. In order to adapt this impressive building, originally designed for agricultural use, it was necessary to redesign its lighting and ventilation, without forgetting the enormous possibilities of making openings due to the excellent situation. The darkest area of the dwelling was illuminated thanks to the courtyard located at the end of the old barn. The beautiful views of the valley and the hills surrounding this house were accentuated by the creation of a large terrace in the east wing of the dwelling and by the creation of new windows. These openings led the architects to create a tortuous and circular layout full of transparencies and unexpected views through the rooms and toward the exterior.

East elevation

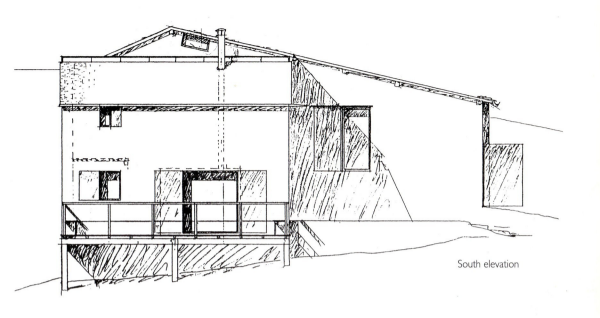

South elevation

On the exterior of the dwelling, with one of its main walls sculpted from bare stone, wood and iron were chosen to follow the style of the agricultural buildings in the area.

Exterior sketch

The excellent geographical location of this dwelling offers incredible views that are accented by the terrace that was built in the east wing and by the large courtyards.

Livingroom sketch

In the large living room, the larch wood combines perfectly with the fireplace built with stones taken from the demolition rubble. A staircases rises above the fireplace, with a small toilet underneath it.

The new volumes are characterised by the type of material used in the furniture: wood, bricks and stone help to define the uses of the different rooms.

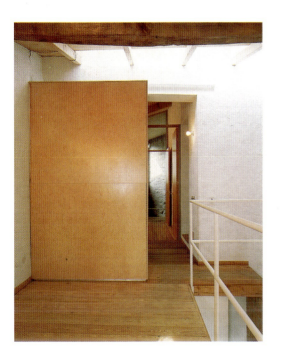

Julian Cowie Architects
Fleetwood Place
London, UK

Photographs: Peter Cook / View

For passers-by, Fleetwood school in Stoke Newington may seem a typical 1880s Victorian institutional building of 1800. However, Julian Cowie Architects and London Wharf developers have changed its interior image completely, successfully converting an undervalued building into deluxe apartments.

In 1917, two large additional blocks were inserted in the angles formed by the main wings of the school. The new floor levels were never aligned and the internal circulation conserves a curious character full of differences. Before one goes through the glazed door, the first impression of the building is one of warmth and comfort. The sandblasted brick walls, the upholstered seats on the landing, and the ash wood doors are small details that give the sensation of great care and elegance in the aesthetic design.

Each of the thirteen floors of the building is different. They all occupy some of the classroom spaces and they share common elements: a double-height space, an inserted kitchen, a central bathroom and a mezzanine (or two) with a bedroom above the service nucleus.

The two flats presented here give an idea of the different inventive transformations that were carried out in these properties.

The first one, located on the upper floor, is occupied by a photographer and has an extraordinary wooden framework that serves as a support for the roof. On this occasion there are two mezzanines which are accessed by means of two spiral staircases of galvanised steel. While one houses the bedroom, the other small one in the centre is used by the owner as a retreat.

He has installed a study below it, partly enclosed by capacious storage walls wich divide the floor space and provide a discrete dining area.

In the second, a music lover occupies a mostly rectangular space in which it was decided to continue the mezzanine along the interior wall in order to provide an additional, fully equipped bedroom. The area of the original mezzanine houses a bedroom and a small corner that is used as an office. The apartments were designed to be flexible, and after a recent visit to Fleetwood, the architect was impressed by the imaginative ways in which the clients had taken advantage of this flexibility.

The main floors have wooden floors. The outer walls of the lower levels conserve the original plaster and windows, but in most areas the brickwork was exposed and the iron beams were painted white.

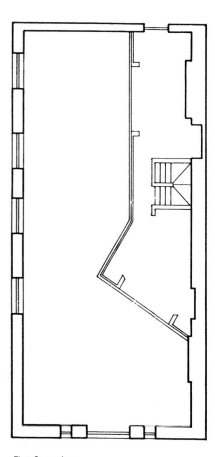

First floor plan

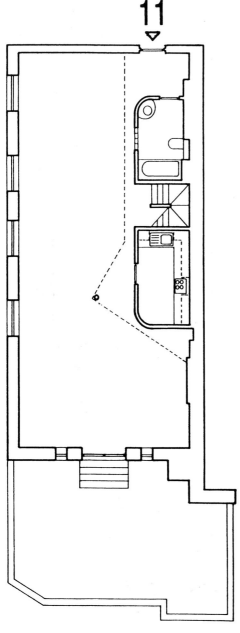

Upper level

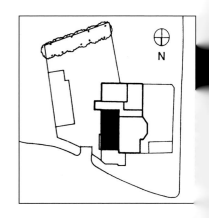

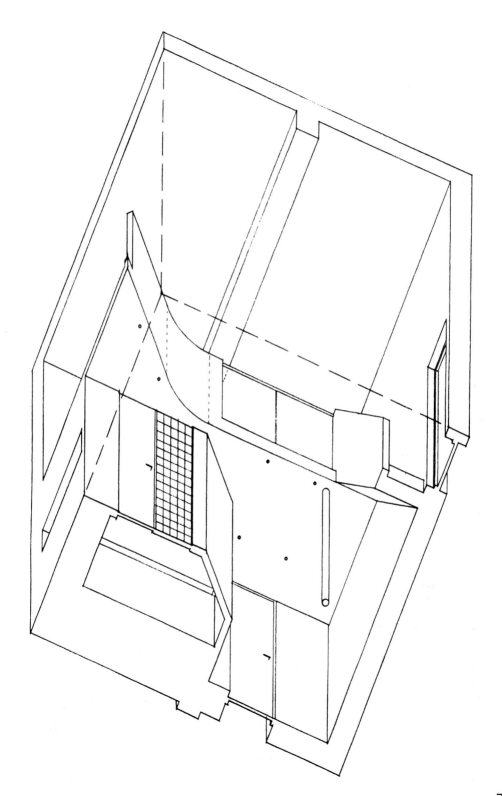

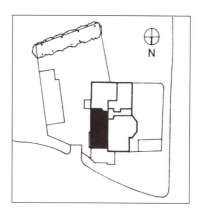

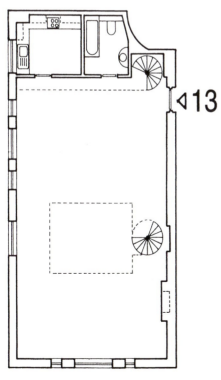

Second floor plan

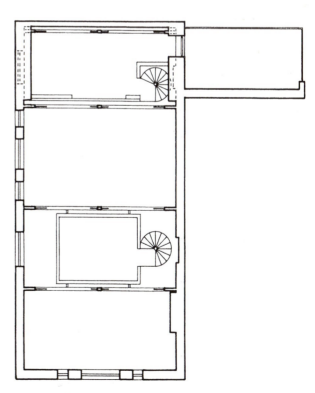

Upper level

Georges Maurios
Rue des Saint-Pères
Paris, France

Photographs: Olivier Wogenscky

For the conversion of this building, the architect aimed to create a privileged space: a group of perfectly equipped apartments in which the small dimensions were no impediment to the creation of a space for working and living comfortably, and in which modernity was no obstacle to maintaining all the characteristics of a traditional habitat.

Due to the conditions of the existing space, the main question for the architect consisted in choosing the best construction method. The brief was to design five apartments of 4.16 m height, although due to the exceptional nature of the location, it was immediately suggested to double the number of spaces by creating duplex apartments. However, the different elements had to be prevented from interfering with each other, bearing in mind the limited floor area and the need to increase the impression of space and seek greater potential use. To achieve this, the duplex system offered the possibility of overlapping spaces with minimum interference, though it called for a careful design to fit the volumes into the dimensions of the building and great care in the construction of the interior.

The spatial relation between the staircase and the office made it possible to create this impression of a vital space on the ground floor, which is all open plan and incorporates a spacious office.

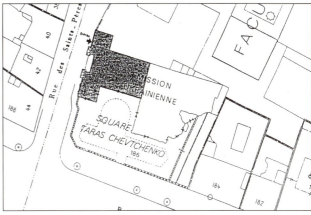

Site plan

The excellent situation of this old property was one of the reasons why it was decided to convert it into a series of apartments of innovative design with all the benefits of modern domestic life.

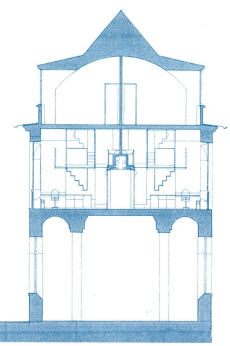

91

Space was saved by the incorporation of a bench for resting by the window and a two-sided articulation block in the kitchen that forms a dining table.

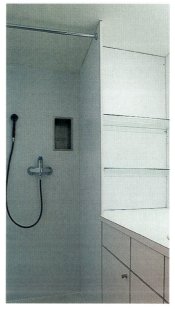

Non Kitch Group bvba
Architecture and lifestyle
Koksijde, Belgium

Photographs: Jan Verlinde

Transforming an old canning factory in Bruges into an impressive loft dwelling was a great challenge with endless attractive possibilities for the designers. One of the most significant features of this scheme by architect Linda Arschoot and designer William Sweetlove, the creators of the Non Kitch Group, was the remodelled roof. It was formed by a lattice structure with a dog-tooth profile supporting a conventional roof. They decided to replace the north sides of each of the parallel roofs with a glazed surface. These skylights greatly increase the natural lighting in the whole dwelling, as in many museums or art galleries. Due to the considerable height of the building (6 metres), this intervention also meant that the interior was almost converted into an outside piazza. A large, full-height room open to the exterior occupies the centre of the space and is surrounded by a mezzanine that houses the kitchen, the dining room, the bar and the television room. Under this mezzanine, three steps below the level of the living room, are the billiards room, the bedroom, the dressing room, the gym and the bathroom that gives directly onto the small garden. The covered pool is located on one side of this exterior space, with an elegant decoration of vertical mosaic strips. An outdoors area also provides better views and enhances the dimensions of the space.

The conservation of the industrial aspect of the building is shown through the use of metal doors, the heating pipes, the separate kitchen, the galvanised iron staircase and the view of the old factory chimney through the parallel skylights of the roof. In opposition to the asceticism of minimalist interiors, the Non Kitch Group feel themselves to be the heirs of the humour and the colourist aesthetics of the Memphis group. One of the premises of this scheme was to generate an appropriate space for viewing the works of art of the private collection of the owners. The furniture is a forceful presence in this dwelling. Designed by Ettore Sottsass, Philippe Starck, Boris Spiek, Jean Nouvel, Norman Foster and the architects themselves, it seems to be made to measure for this spacious loft.

Architekturbüro Gasparin & Meier

Badehaus Ebenberger
Siflitz, Austria

Photographs: Margherita Spiluttini

Located 1000 metres above sea level, this old two-storey stone farmhouse was a damp place without running water that was urgently in need of restoration. The absence of running water aroused the desire to create a special area for the bathrooms without spoiling the open plan of the existing structure.

Access to the building is through a striking large entrance located in the centre of a stone wall. This door gives access to two large rooms located on both sides of this entrance that are used as bedrooms.

In the essential restructuring, the tarmac road that goes down the mountain had to be moved because it was located too near the building. The foundations were left free to protect the walls from damp. The wooden bathroom annexe located to the west is now the sunniest area in this solid stone house.

This room, which was formerly used for smoking bacon, now houses the bathroom with its washbasin, bathtub and other complements. It is a spectacular space which also gives access to a corridor 8 metres long 1.5 metres wide that is used as a luxurious sauna and as a bridge that communicates with the old area of the building. One of the most striking aspects of this scheme is the importance given to the bathroom, which is further enhanced by the large sliding window that offers excellent views of the valley. Another of the outstanding features is the use of stone cut from the higher part of the mountain for the pavement and part of the walls.

Under the old stone wall, which had to be demolished and reconstructed, a new basement was created taking advantage of the old staircase. It receives light through the glass at the bottom of the bridge.

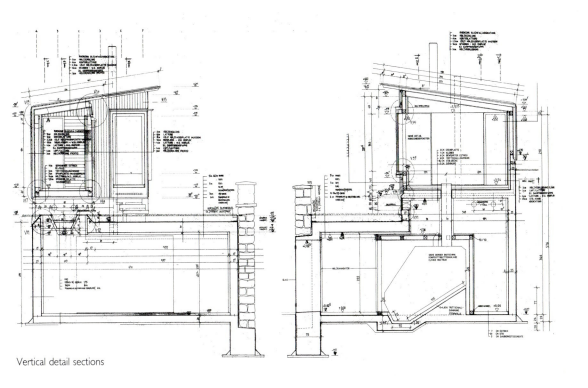

Vertical detail sections

The new bathroom gives access to a corridor 8 metres long and 1.5 width metres wide that is used as a sauna and serves as an access bridge to the old area of the building.

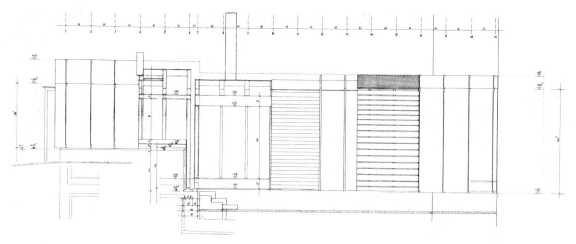

Side elevation

A new wooden annex in the west area serves as a bathroom and sauna for an old farmhouse. This intervention also improves the natural lighting of the house.

Patrick Genard & Ass.
House T
Barcelona, Spain

Photographs: Eugeni Pons

The rehabilitation of this tower with a garden involved an interesting exercise of transmutation of polarities: tradition-modernity; culture-nature; minimalism-baroque style…

The final result is therefore the product of these opposing elements, but also of their transcendence in the search for an unstable point of balance. The volumes and composition of the facade were entirely conserved by the architects due to the planning regulations, but they treated it graphically and in two rather than three dimensions, replacing the relief elements with recessed joints. In the interior, the volumes were also respected but given fluidity by all types of visual transparencies. A space of double height was thus opened at the entrance. The stairwell was eliminated, opening many dynamic perspectives, and transparent partitions were created for continuity of the spaces. The ceilings were separated from the walls by means of an illuminated throat to blur the limits. The architects removed all structural elements in the space located under the roof, thus forming a large suite in the maisonette, and the transparent character of the rooms was enhanced by means of the repeated use of doors, sliding walls and spiral staircases of translucent glass. The architects also attempted to merge the minimalism of the layout with a certain "baroque style" in the natural materials, chosen for their strong personality and their great decorative power.

This whole endeavour is sustained though great care in the geometry and proportions and a harmonic composition of ranges of materials, from the furniture and the details to the construction itself and the works of art. These were chosen during the project in an interesting dialogue with its aesthetics, following the same philosophical line.

In summary, it was attempted to integrate the polarities to reach a harmonic and organic whole, to conjugate the objects in a whole in which opposites approach and resemble each other.

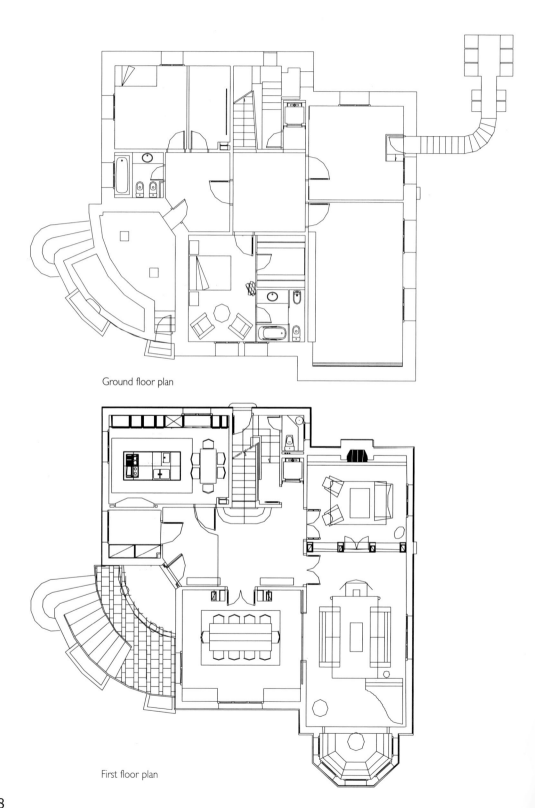

Ground floor plan

First floor plan

In the rehabilitation of this dwelling, one of the main aims was to create different spatial atmospheres in which the continuity was not interrupted by too many visual barriers. In the living room, a translucent partition and a sliding glass door that communicates with the dining room help the natural light to filter toward the interior.

ARTEC Architekten
Raum Zita Kern
Raasdorf, Austria

Photographs: Margherita Spiluttini

Following the needs of the owner, a literature scholar who was fond of the country life, a new and provocative building element was constructed in the old farm of Marchfeld. The fundamental requirement was to create a large and comfortable office for reading and writing, far from the noisy atmosphere of the city. The conversion also required the construction of a bathroom and a toilet and the installation of a complete heating system.

The starting point for the design was the former cowshed, whose roof was in danger of falling in and had to be completely removed. The new roof was prolonged with the incorporation of a staircase located on the exterior, creating an unusual shape that marks the contrast between old and new. It was attempted to take advantage of the original module as far as possible, so that the same building served as a basis for a new architectural element. The whole facade of the cowshed is clad in aluminium, creating an experience that remains in the memory of its visitors. In the interior, materials such as poplar plywood and aluminium move away from the rural atmosphere and provide a cosmopolitan and welcoming air.

The bathroom is located in the main building and is illuminated by a flat skylight. On it, the rainwater can collect up to a depth of two centimetres, thus creating special effects of shades.

The new office is located on the first floor, a space that is calm and distant from its surroundings and conducive to the full force of the intellect. This study gives onto two terraces through sliding doors that extend the room even more. The location of the terraces helps to illuminate the interior of the dwelling and makes a decisive contribution to the peculiar form of the scheme.

The opposition between the construction materials of the old building and those used for the new room creates a special effect by merging the urban and rural atmospheres.

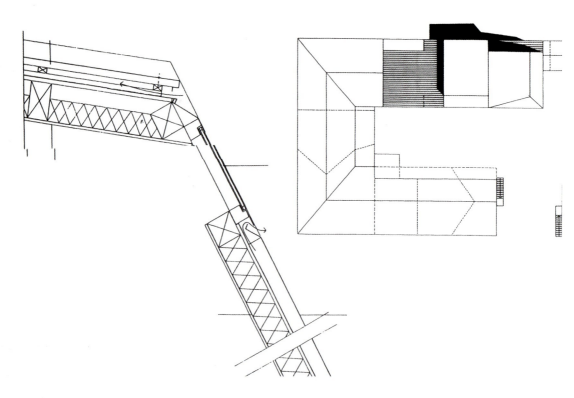

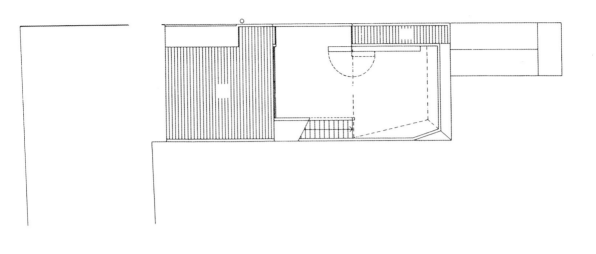
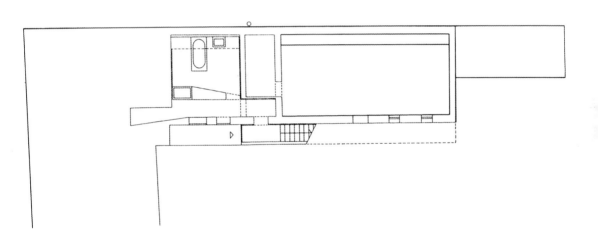
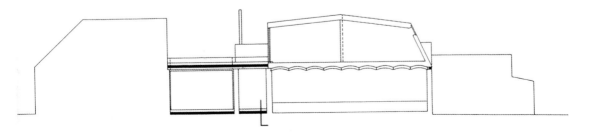

Brightness and simple forms were key elements in defining the interior spaces of this dwelling. This is most clearly appreciated in the colour and the smooth textures of the staircase and bathroom.

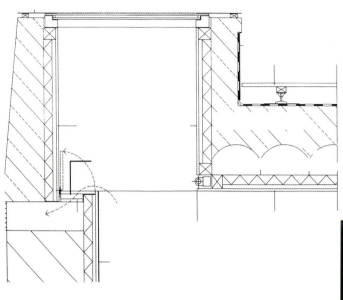

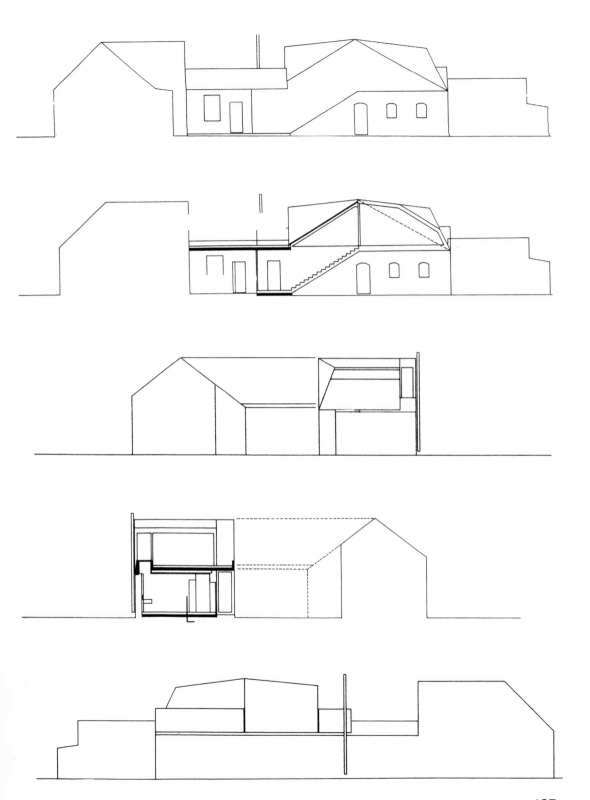

Klaus Sill & Jochen Keim
Apartment and Office Building
Rathenow, Germany

Photographs: Christof Gebler & Klaus Sill

The building converted by Klaus Sill and Jochen Keim was built one century ago, although it was later modified by joining the front area to the courtyard. On this occasion, the building has been integrated homogeneously in an environment that is distinguished by the walls of the block and the intelligent conception that makes it suitable for working and living.

The courtyard building had fallen into a state of ruin in the few last decades: some walls had collapsed and others were in danger of collapsing.

It was decided to only conserve an old dwelling and a warehouse, two elements located at the north-west end of the land that were considered to be worth maintaining.

As part of the project, a major installation was built inside the block to serve as a landscaped area and playground for the tenants of the front building. Also, thanks to this arrangement, it was possible to cover the objectives agreed in the urban plan of the city of Rathenow concerning the separation between the blocks and the foundations.

The conjunction between offices and dwellings was achieved by means of a distribution in which the upper floors were reserved for the dwellings. Thus, the ground floor and first floor contain the cubes (12 modules) that house the offices of a firm of engineers with twenty employees. The second floor is composed of three maisonettes of about 60 and 90 sqm.

For the firm of engineers two united surfaces were used because the building protrudes about 4.5 m into the courtyard. This extension is formed by two conferences rooms, an audio-visual room, an archive, a kitchen and a toilet.

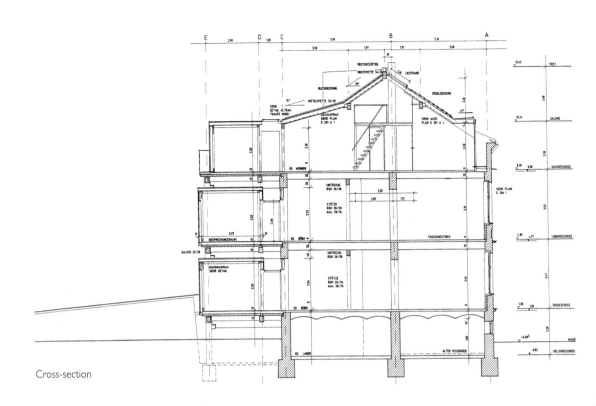

Cross-section

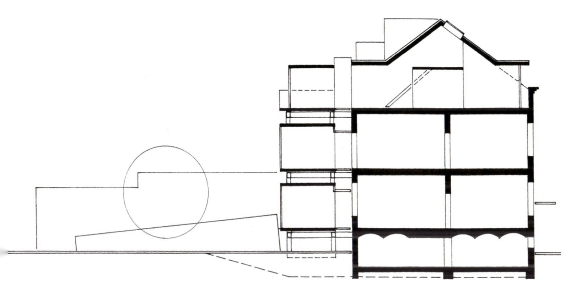

Cross-section

The project takes all its characteristics from the building and from the environment in which it is located, and it integrates the mixture of dwelling and workplace in a way that is more in line with the current lifestyle.

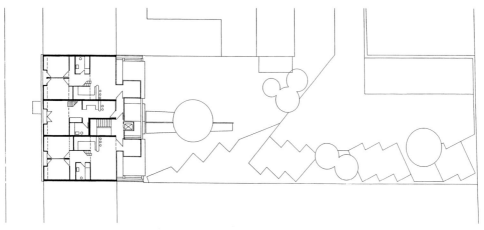

Attic

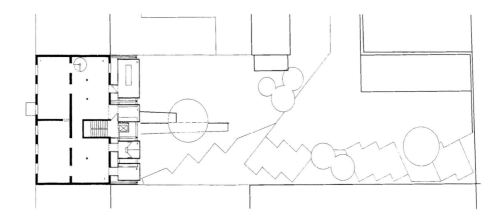

Second floor plan

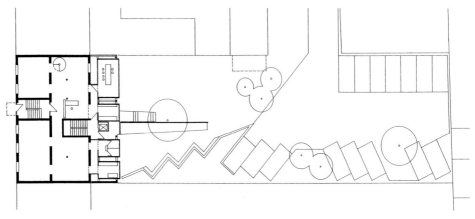

First floor plan

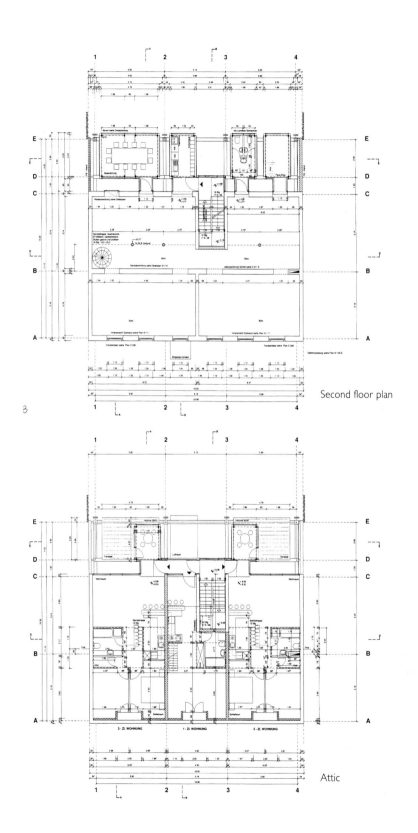

Second floor plan

Attic

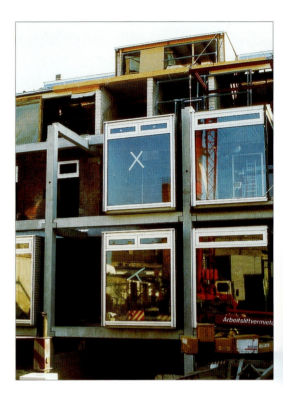

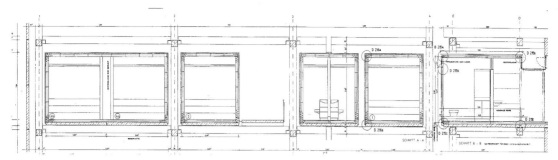

Section A-A B-B

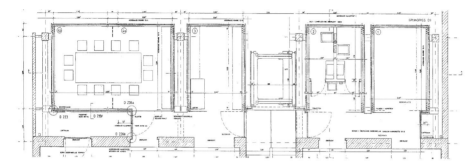

Ground floor plan

Axonometric view

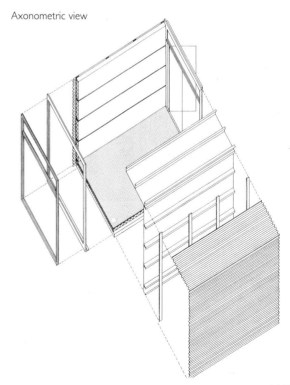

or economic and technical reasons, these modules were built 500 km from the site because they were of a suitable size to be transported. They were set into a prefabricated reinforced concrete skeleton for fire protection.

Jo Crepain Architect NV
Water-tower
Brasschaat, Belgium

Photographs: Sven Everaert / Ludo Noël

Until 1937, this water tower with a height of over 23 metres was used to provide water to the castle and other buildings of the county of Brasschat, near the city of Antwerp. After being in disuse for decades due to the construction of four new water tanks and the planning of a modern water supply system, it survived a planned demolition. The conservation of this peculiar cylindrical tower crowned by a large, four-metre-high cistern allowed it to be converted into an unusual single-family dwelling. The architect respected the original industrial typology, leaving the four large pillars that sustain the structure exposed, and also maintained the compositional structure and the essential functionality of the original design. This was achieved by minimising the presence of decorative objects and by limiting the elements and materials to reinforced concrete, structural glass and galvanised metal. Around the original structure, a parallelepiped, double-height volume with a mezzanine surrounds the tower at ground level. This new construction houses the services and a living room that is totally open and transparent to the exterior. This breaks the verticality of the scheme and gains space, and its roof acts as a terrace for the first floor, which houses the main bedroom.

The new tower achieves its maximum expressiveness when it is illuminated at night. The transparency of the glass structure that wraps the building allows the occupants to enjoy the wooded landscape with a small winding creek and reveals the three floors of 4x4 m, each with a small balcony. These floors house, from bottom to top, the study, the guest bedroom and a small winter garden. At the top of the tower the water cistern is conserved, now transformed into a curious space without windows that is intended for private receptions.

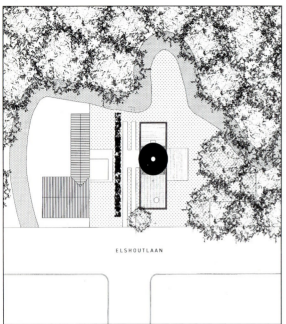

Site plan

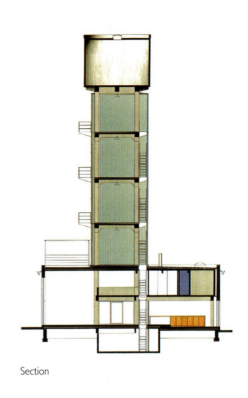

Section

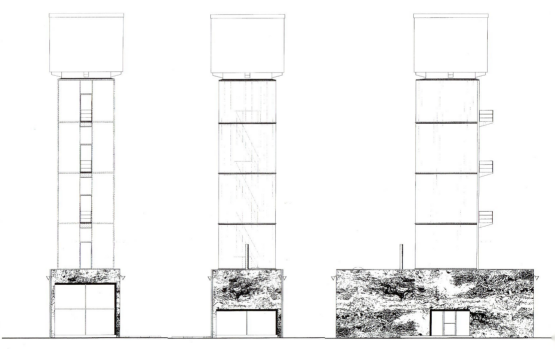

Cristian Cirici & Carles Bassó
Vapor Llull
Barcelona, Spain

Photographs: Rafael Vargas

The Vapor Llull (a steam-driven factory), in an old industrial district of Barcelona, consisted of a set of buildings dating from the early 20th century which had been devoted to manufacturing chemical products.

The basic structure of the complex consisted of a long ground floor plus two floors, the highest of which had a sloping roof supported by a structure of wooden trusses. The complex also included a series of auxiliary premises adjoining the main building and a magnificent brick chimney over thirty metres high that was part of the steam engine that powered the factory. The architects decided to conserve the chimney in order to keep alive a symbol of a time in which the whole district was full of steam-driven factories. The most suitable property for conversion into loft dwellings was the long main building. In order to create an open, private space and provide a one-vehicle parking space for each of the eighteen units into which the scheme was subdivided, a series of auxiliary buildings were demolished. To give independent access to each module of approximately 90 square meters, three sets of vertical communication elements were introduced, each with a stairwell and a panoramic elevator. Their formal expression gives the appearance of silos covered with enamelled corrugated steel. On the outside the main building was painted with silicate paint applied directly to the bricks, which were first stripped of their render.

In the interior, it was decided to leave the spaces free and unfinished, so that each loft could be arranged according to the wishes of the different designers of interiors that were chosen to finish off the scheme. The layout and decor of this loft are by Inés Rodríguez. It is an two-level apartment in which a mezzanine houses the bedroom and a bathtub. It is a curious habitat in which the light and the space create an atmosphere of elegance.

The sophistication of the newly-built elements contrasts with the industrial character that the building conserves. The colours, and the use of large numbers and letters on the facade and doors, bring contemporary architecture towards graphic design.

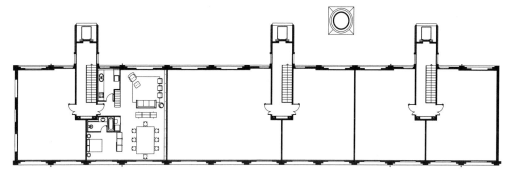

Second floor plan

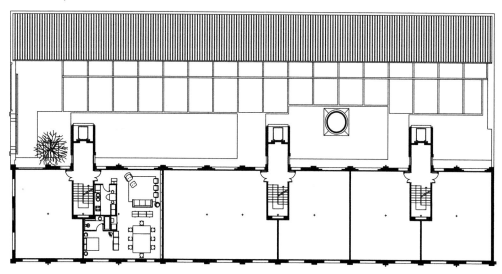

First floor plan

Ground floor plan

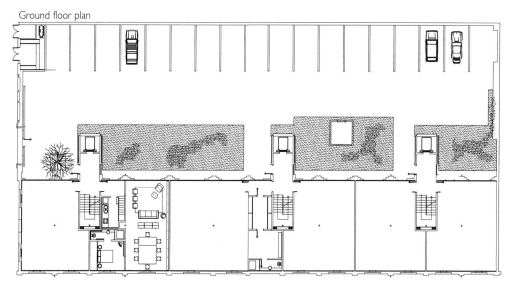

Panoramic detail of elevator windows

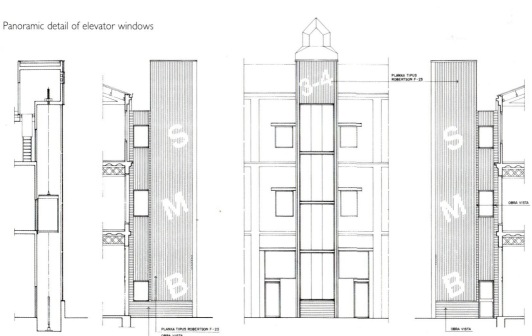

In this apartment large spaces with bright, clear walls are predominant. The most outstanding features at first sight are the wooden trusses in the whole dwelling and the polished concrete floor.

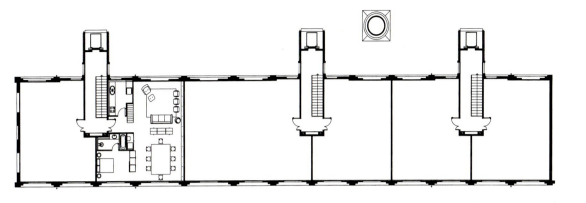

Fernando Távora
House in Pardelhas
Pardelhas, Portugal

Photographs: *Luis Ferreira Alves*

This country house is located in a mountainous area near Vila Nova de Cerveira, in northern Portugal. It is a farm building with spaces of small dimensions that forms part of a set of three houses with identical typologies, each one defined by a closed exterior space. The three different nuclei were practically in ruins and their structure was hardly supported by the wide granite walls.

The houses establish relationships of vicinity so that different exterior spaces enhance each other and give form to the whole, with enveloping walls and corridors.

The scheme seeks to take advantage of the habitability of the existing structure, adapting it to a new use and recovering the traditional construction process of tiled roofs with wooden beams that provides a better reading of the existing structure. The new layout is freed from the upper space to give way to the main room, which is in direct contact with the kitchen, and to a second room that was entirely rebuilt in wood. The private area of the house, containing the bedrooms and bathrooms, is located on the lower floor in direct contact with the exterior courtyard.

The canopy of the entrance passage is made of wood, supported by existing stone platforms, establishing a continuity with the exterior railing and with the main room. The house and the annex are joined by a corridor, defining a more private and independent space.

The house itself follows the same logic of intervention, using the walls to contain its new spaces, which form an independent section that is self-sufficient from the main house, with bedroom, living room, kitchen and bathroom.

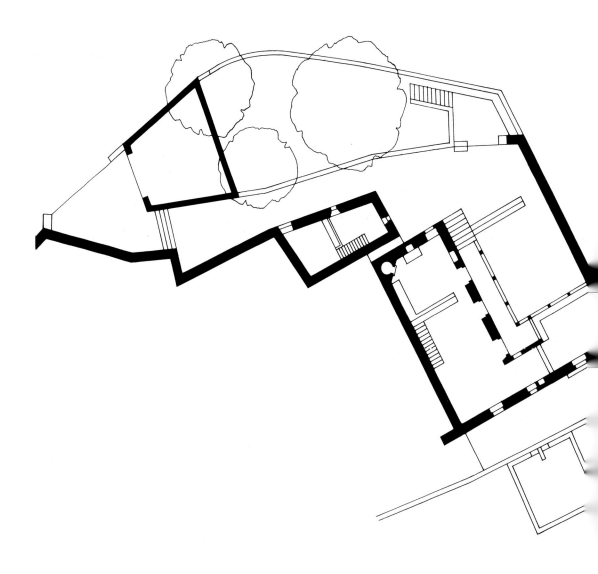

The walls are finished with stone blocks, following the style of the mountain houses in this area of the River Miño, giving the house a distinctive vernacular touch.

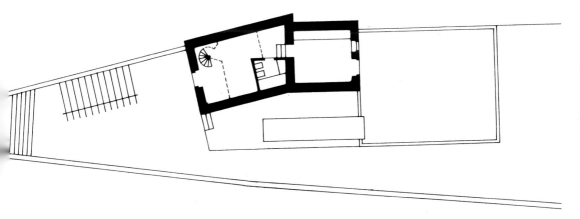

Site plan

The rural character of this dwelling and its environment disappears almost completely in the interior. The combination of new materials helps to emphasise the natural lighting, creating an unexpected feeling of comfort in a house built of granite.

Jean-Pierre Lévêque
House Rue Compans
Paris, France

Photographs: Hervé Abbadie

This building, an old laboratory built in the thirties, was transformed into an inhabitable space after twenty years. It is located in a complex fabric of plots, a kind of a residual space in the form of an isosceles triangle between two five-storey blocks.

For the rehabilitation of this small dwelling of 80 sqm, the brief was to optimise its habitability and to rediscover in the space a clear legibility and its initial constitution as a building suspended over a covered exterior.

The layout offered the possibility of creating a multi-purpose space defined by the exterior and interior elements. The ground floor was left completely open in order to allow the exterior, consisting of the covered courtyard, to be extended completely into the house. Inside this covered exterior, a differentiated structure containing the kitchen was inserted. This "box" is completely open within the continuous layout of the floor, walls and ceiling, thus giving the dwelling a distinguished appearance.

The exterior of this volume is secured by means of the lower part of the room, and by the pillar that supports it. The house is therefore suspended, with the areas that require greater domestic privacy, such as the bedrooms and the bathroom, at the top.

All the spaces were connected to each other by means of a wooden strip. This begins as the main envelope of the kitchen, becomes the staircase that gives access to the first floor and ends in a wide bookcase before leading to the bedrooms. This "Ariadne's thread", ensures maximum fluidity, multiplying all possible points of view on the area of the dining room and the kitchen in the ascent to the first level, in which all the dimensions of the space that one moves through are apparent.

The basement is accessed by a long flight of stairs that is partially covered by a glass panel. This opening provides natural lighting for the office located in the basement, and accentuates the effect of inclusion of the kitchen volume in the ground floor.

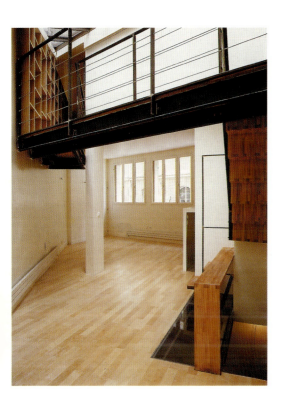

The walkway -which serves as an access to the main room -maintains a continuity with the staircase and the kitchen module.

Claesson, Koivisto, Rune

Private apartment / Town house
Stockholm, Sweden

Photographs: Partrick Engquist - Åke E: Son Lindman

The team of architects Marten Claesson, Eero Koivisto and Ola Rune were contracted to restore two apartments in the centre of Stockholm.

In the first one, for a young manager with little free time, they attempted to organise a peaceful home for "charging the batteries" after long working days and trips. This dwelling is located in a building of the late 19th century with a wooden floor. It was decided to maintain one half original -the living room- and to make all the other rooms -the bedroom, kitchen and bathroom- totally modern. In order to further emphasise the contrast between the two areas, the new area was painted pale grey and the original area was painted white. The floor plan was devised in order to create visual fluidity between the rooms. To communicate the spaces in a simple and functional way, two new structures were designed. A curved corner leads from the entrance into the first axis along bathroom, kitchen and bedroom. The second axis of intersection runs from the kitchen to the dining room, where a wall with a hatch was built.

The final result was a comfortable apartment full of light, in which the arrangement of the glass panels and the choice of furniture were the essential elements for the architects.

The second scheme was set in a Neo-Classical building of the early 20th century. The client acquired the whole property and decided to remodel two of the flats to create his private dwelling: a large, bright apartment giving onto the garden.

The spaces had previously been used as small dwellings and had undergone many modifications over the years. To create the new dwelling, most of the walls were demolished, while the floors, the windows and the original radiators were conserved. One of the main interventions was the creation of a complex stairwell between the two floors, with glazed openings like those of churches. On the upper floor a major feature is the modern design of the bathroom, with a high bathtub designed by the architects and placed strategically to offer panoramic views of the port of Stockholm.

Private apartment

Town house

The dishwasher, refrigerator, freezer and oven are totally integrated behind the doors. Great care was taken with the lighting and the choice of furniture.

An essential factor is the vertical and horizontal composition of the openings: a vertical opaque pane in the door communicating the entrance area and the bathroom, a horizontal one between the bathroom and the kitchen, a translucent vertical pane between the dining room and the bedroom, and a mirror at the end of the bedroom wall creating the illusion of a continuous space.

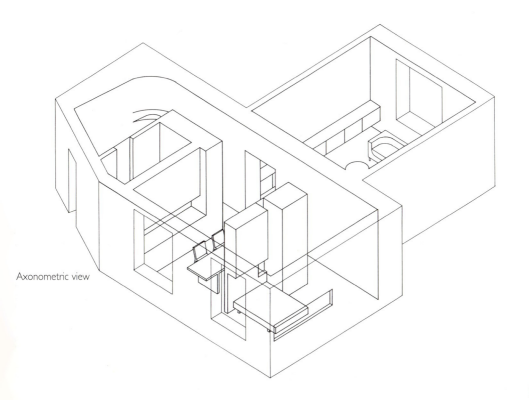

Axonometric view

193

Town House

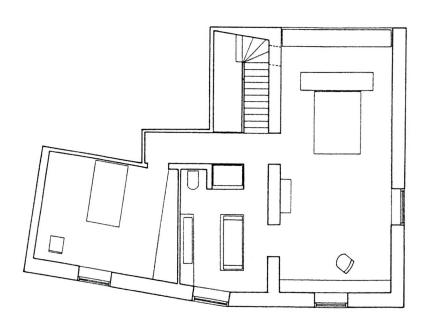

The geometry of this project is distinguished by its simplicity and functionalism. The straight lines and the contrast between the different materials highlight the new volumes, giving the whole a modern appearance.

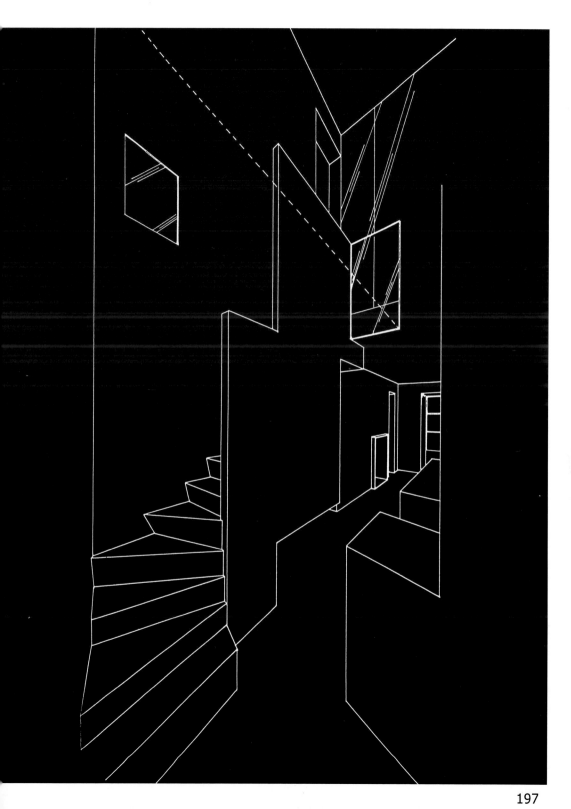

Littow architectes

Paris 6e

Paris, France

Photographs: Pekka Littow

This apartment occupies the last two floors of a Parisian property built in the 17th century. These last two floors were probably built more recently to house two simpler apartments for the servants. It was later decided to join them to form a single apartment. After a meticulous study of the technical limits of this conversion, the challenge consisted in obtaining a space that was as open as possible. All the existing elements that limited the views or the natural lighting were reduced without sacrificing the hierarchy of the spaces and the structure of the inhabitable areas. It was thus attempted to create subtle and delicate borders that could be adjusted. They are like visual borders that allow for different interpretations. The dining room is isolated from the other rooms by glazed walls, allowing in the natural light from the inner courtyard and from the openings in the guest room. The kitchen forms an integral part of the living room, although it is completely concealed by a set of folding wooden panels, so the appearance of the apartment varies according to the time of day and the needs of its occupants.

The abundance of exposed pillars and wooden props, many of which were completely rotten, called for special treatment and some were eliminated or replaced. The brick walls of the street facade were stripped and according to age-old techniques, showing a craftsman's touch that gives new value to the building.

In the upper room the drop ceilings were removed to reveal magnificent woodwork. The same procedure was applied to the guest room, where a small mezzanine was installed to the right of the volume of the old covered structure. Inside the dwelling, the new spaces that have been created reveal more about the demolition and the subtle unification than about the construction itself. Composing with the existing elements was a priority, but the question was to define what could be changed, what had to be conserved and how to deal with it. The first rule was that the original structures should be the dominant ones, the new elements only playing a secondary role.

The furniture was designed by the architect and made specially for these spaces with the aim of creating a relaxing atmosphere.

The undulating oak ceiling gives beauty to the room and provides excellent acoustic insulation.

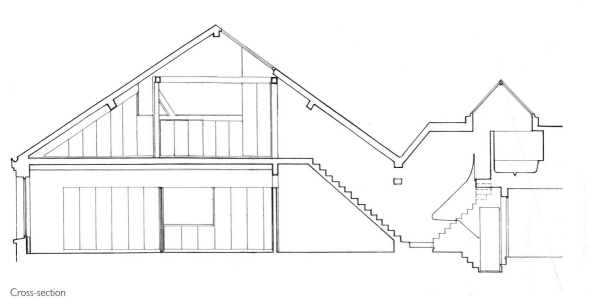

Cross-section

203

The creation of a bedroom under the beams of the loft is a practical solution that gains space that would be normally remain unused. With this option it is also possible to create a different atmosphere in which the mixture of materials forms an essential part of the design.

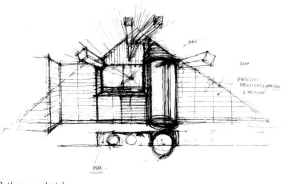

Bathroom sketch

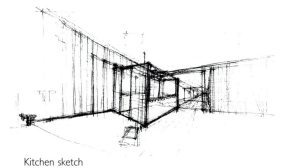

Kitchen sketch

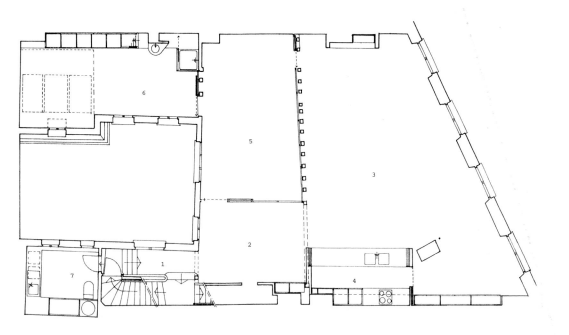

First floor plan

1. Entrance
2. Hall
3. Living room
4. Kitchen
5. Dining room
6. Guestroom
7. Laundry

Upper floor plan

8. Master bedroom
9. Bathroom
10. Toilet
11. Mezzanine

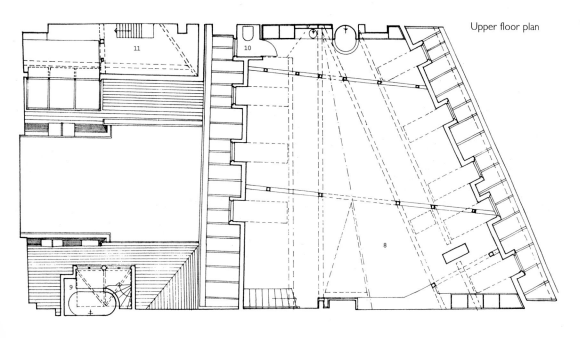

Upper floor plan

José Gigante
Wind Mill Reconverting
Vilar de Mouros / Caminha, Portugal

Photographs: Luís Ferreira Alves

In the grounds of a recovered house in northern Portugal, an old abandoned windmill waited its turn to be useful again. In the course of time the idea finally arose of transforming this peculiar building into a small auxiliary dwelling belonging to the main house, giving it its own life and thus creating a completely inhabitable and independent space that could be used as a place of rest. For José Gigante, the architect in charge of the conversion, the presence of the mill was so strong that any major intervention would have minimised its charm. Therefore, without touching any of the thick granite walls, an unusual cooper roof with a very gentle slope was added. The intention was to respect the memory of the place as far as possible, so the inspiration for the transformation began naturally from the inside towards the outside. The layout and organisation of the small space, with only eight square metres per floor, was not easy. Thanks to the choice of wood as the main building material, a welcoming atmosphere enhanced by the curved walls and the few openings was achieved. On the lower floor, an impressive rock acts as an entrance step. On this level it was attempted to achieve a minimum space in which it was possible to carry out different activities. It houses a bathroom and a living room, with the possibility of transforming a small sofa into a curious bed: it is conceived as a case that contains all the necessary pieces for assembling the bed. On the upper floor, the furnishings are limited to a cupboard and a table/bed that is extended to the window.

The only openings are those that already existed in the mill and they have been left as they were conceived, with their natural capacity to reveal the exterior and to illuminate a space in which the contrasts between the materials cannot be ignored. The typology of this building was crucial to the restorations to which it has been subjected, and shows why the interior space is so important in this scheme. The thick circular walls occupy more space than the interior of the mill, but they hug the whole room and provide a welcoming and unconventional sensation that give this building a new and innovative perspective.

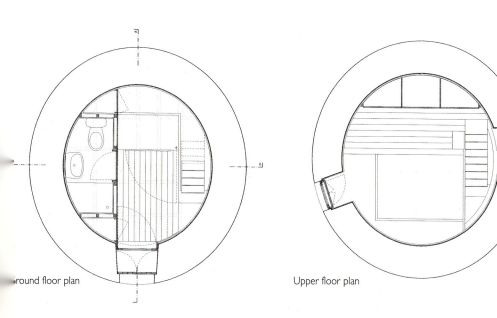

Ground floor plan　　　　　　　　Upper floor plan

The choice of wood as the main construction element in the rehabilitation, in perfect combination with the white curved walls, creates a calm and welcoming atmosphere.

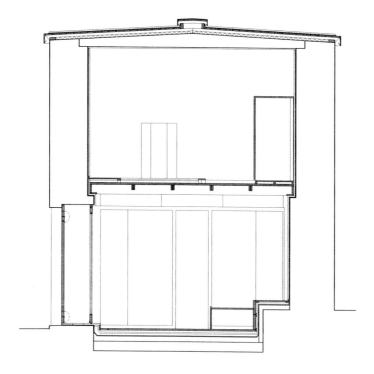

Cross sections

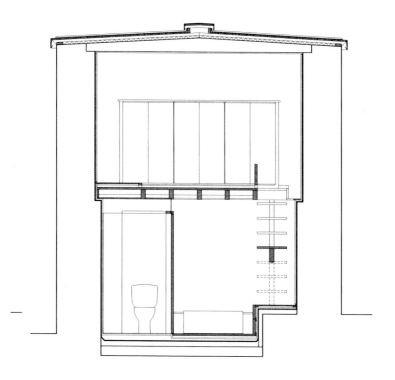

The use of wood and the maximum use of space are the main characteristics of this scheme. To solve the problem of the lack of space, a system was devised in which a bed is hidden at the foot of the staircase.

Studio Archea
House in Costa San Giorgio
Florence, Italy

Photographs: Alessandro Ciampi

This apartment designed by Studio Archea is located in an old medieval tower near the Ponte Vecchio, in the heart of Florence. The original Renaissance building had large wooden beams that gave it a certain majesty, and the challenge consisted in designing a residential space that took advantage of the exceptional characteristics of the quattrocento to create a functional and contemporary atmosphere.

The space is designed around a curvilinear stone wall that organises the diverse functions of the dwelling and supports the metal beams of the mezzanine, which is used as the night area. This wall acts as a bookcase, leaves the kitchen semi-concealed, houses in its perimeter the stairwell, and divides the spaces so that the materials define and organise the different atmospheres of the dwelling. An iron staircase set against the opposite wall leads to a platform giving access to the mezzanine. From this horizontal platform, a walkway also leads to a small panoramic pool over the dining room with a bathroom next to it. This area of the upper floor is in the new stone part of the apartment; the rectilinear mezzanine is separated by a small wooden floor space. Because of the small size of the scheme, the architects were able to design all the elements in detail, avoiding prefabrication and creating unique, almost sculptural objects.

The architects showed great respect for tradition in the use of natural stone, in the conservation of the original ceiling and in the distribution of the furniture. A single space and different atmospheres for one person: this is the result of this intervention in an apartment set between walls full of history.

The void created between the two main volumes is used as a corridor on the ground floor. On the upper floor, this free space offers different views and perspectives, and opens up the dimensions of the apartment.

Aneta Bulant Kamenova & Klaus Walizer
Conversion Sailer House
Salzburg, Austria

Photographs: Rupert Steiner / Archiv Eckelt-Glas

The owners of this villa in Salzburg are art collectors who wished to add spaces connecting the house to the garden that were protected from the weather and could be used throughout the year. This brief seemed compatible with the architect's idea to use techniques for ecological buildings that reduce climatic differences. After eliminating all the unnecessary building elements that were added in the fifties, the house's form of a closed prism reappeared. On the north side, a garage for two cars and an entrance like a small porch of satin-finish glass were added, both connected with an overhanging glass roof of 1.6 m. The main interest of the architects was to open the house toward the garden located in the south. To achieve this, they created a large terrace in which the glass-covered garden meets a pergola with tensioned cables that provides shade through the plants. The key to this spatial ensemble was to obtain a passage to the exterior through different climatic areas, and also to create a "membrane space". One can thus feel the metamorphosis of nature at all times of the year and enjoy a beautiful sensation under the rain of Salzburg, since the reflected images of the garden seem like magic events.

The building is therefore transformed into an activating organ for perceiving the changes of nature. A special characteristic of this winter garden is the construction technology. The glass cube has a skin of insulating glass and a glued glued glass-only construction. The basic structure consists of two columns, screwed to two beams at the front side, constructed in triple-sheet laminated glass. The roof boards of isolated toughened safety glass with "fritted" ceramic patterns, casting 40% shadows, are glued to these beams. The minimal number of screwed connections is not part of the carrying structure, but has been caused by the necessity of faster drying-time of the silicon-adhesive (as the building was constructed in winter). Another innovation is the frameless door of insulating glass, which is the first of its kind. This type of construction in which the glass is secured by adhesive is a technological innovation that had never been used in Austria, where the climate is quite harsh. The use of a single material, glass, and the simple details of the structure give this high-tech assembly the appearance of a simple construction. Through this paradox the architects avoid the effect of "technical expressionism".

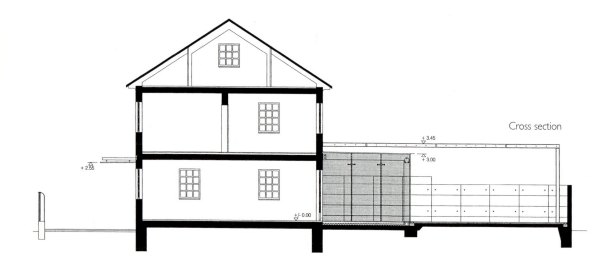

Cross section

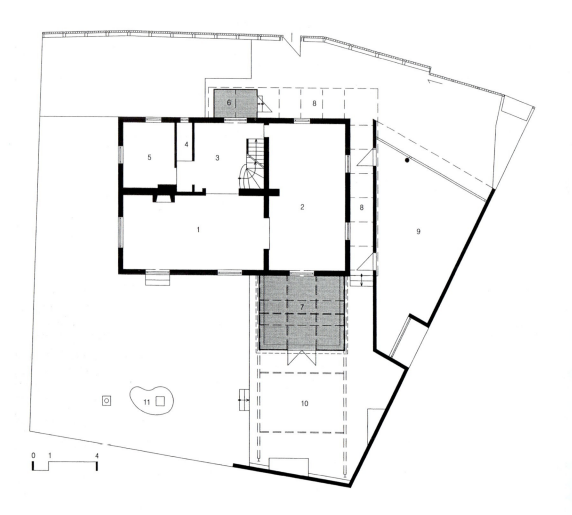

Cross section

1. Living-dinning-room
2. Musicroom
3. Entrance hall
4. Bathroom
5. Kitchen
6. Porch
7. Winter garden
8. Glass roof
9. Garage
10. Terrace with pergola
11. Fountain with sculpture

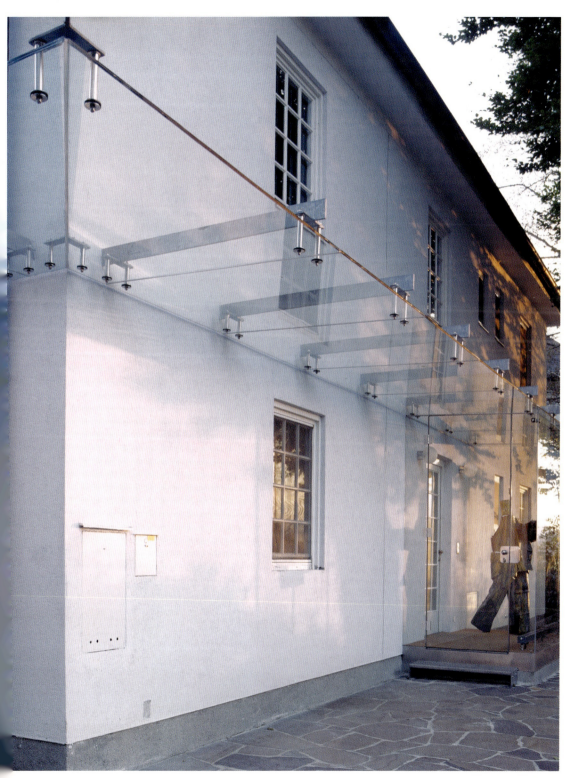

Fraser Brown MacKenna
Derbyshire St. Residence
London, UK

Photographs: Nick Hufton

Following personal recommendations, Fraser Brown MacKenna Architects were chosen by the client to design a new house in an abandoned shopping area of Hackney. Although finally the site was not purchased, the plans for the scheme were well received and included in the RIBA'S "First Sight" exhibition in 1995. The client then asked Fraser Brown MacKenna to restore the two upper floors of a converted loft in Bethnal Green. A dramatic barrel vaulted Perspex roof had been added to the building but spaces within were subdivided and darked.

A translucent floor was used in a space of double height, the upper floor being used as the main area with the kitchen and dining room. The lower floor was stripped and opened to achieve a large flexible level for sleeping, working and living. On this floor, a ten meter long, full height Plexiglas screen, divides storage and utility spaces from the main circulation and living areas in the building. It provides a bright, clean plane in contrast with the large expanse of the restored floorboards. The light is reflected in it, causing effects of brightness and shade that give the impression that the light comes alive as one moves. This screen can be manipulated to modify the space, thus creating a new area for sleeping, talking or relaxing and allowing the space to adapt to the needs of the owners. The screen employs a bold language of fixing that are expressed in the stainless steel socket cap head screws and aluminium T-sections. Against the screen, the simple geometry of the copper clad of the storage area reveals within its natural patina a hundred hues to play against the subtle shadows of light. The simple form of screen, of box, of sink and fin walls formed from glass blocks or perforated metal sits within acknowledges the space and what it has been before.

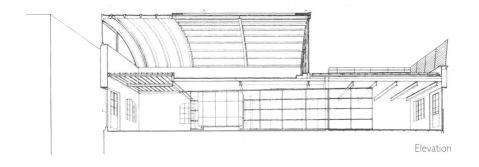

Elevation

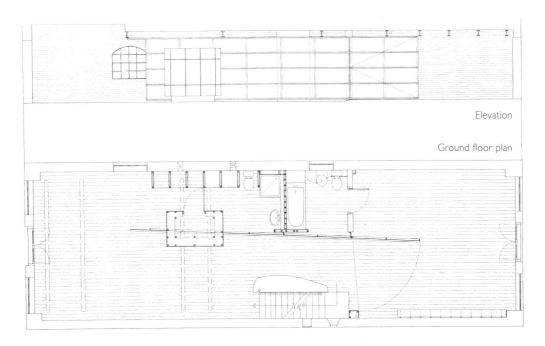

Elevation

Ground floor plan

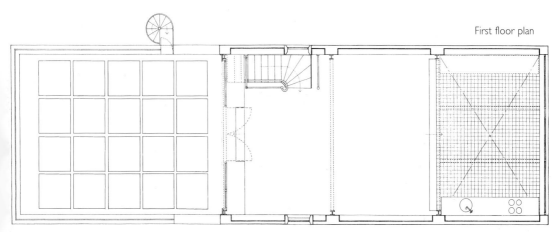

First floor plan

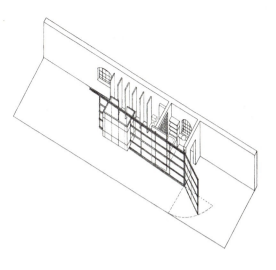
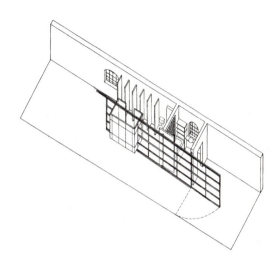

A Plexiglas screen divides and modifies the space according to the needs of its owners. The material gives a modern touch and the light filtering through it illuminates the whole apartment.

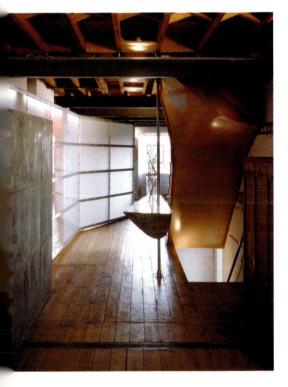

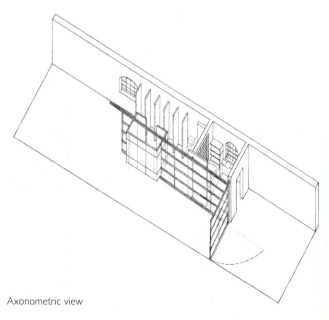

Axonometric view